FINDING
your
TREASURE

FINDING *your* TREASURE

OUR FAMILY'S MISSION *to* RECYCLE,
REUSE, *and* GIVE BACK EVERYTHING—
and HOW YOU CAN TOO

ANGEL WILLIAMS

Tiller Press

New York   London   Toronto   Sydney   New Delhi

TILLER PRESS

An Imprint of Simon & Schuster, Inc.
1230 Avenue of the Americas
New York, NY 10020

First Tiller Press hardcover edition July 2021

TILLER PRESS and colophon are registered
trademarks of Simon & Schuster, Inc.

For information about special discounts for bulk purchases,
please contact Simon & Schuster Special Sales at 1-866-506-1949
or business@simonandschuster.com.

The Simon & Schuster Speakers Bureau can bring authors to your live event.
For more information or to book an event, contact the Simon & Schuster Speakers
Bureau at 1-866-248-3049 or visit our website at www.simonspeakers.com.

Interior design by Laura Levatino

Manufactured in the United States of America

1   3   5   7   9   10   8   6   4   2

Library of Congress Cataloging-in-Publication Data has been applied for.

ISBN 978-1-9821-5229-1
ISBN 978-1-9821-5230-7 (ebook)

All photographs courtesy of the author.

This book is dedicated to my grandmother Melvirdia Shelton.

Grandma is no longer with us, but she will always

hold a special place in my heart. ❤

Contents

FINDING *your* TREASURE

Introduction

I 'll begin this book the same way I begin all my YouTube videos: "Good morning, everybody!"

In 2012 I started posting my dumpster-diving adventures on a YouTube channel called Mom The Ebayer. It's hard to believe, but it now has over 150,000 subscribers. If you've never tuned in, I'm a spirited—and spiritual—mother of four. I started dumpster diving in and around Chicago after a parishioner at my church mentioned scoring some surprising finds. My curiosity got the best of me (it often does), and so I gave it a go. On my very first outing, I returned home with a mint-condition playground set for my then eighteen-month-old. Total cost: zero dollars. Not to mention a BlackBerry that I then sold for $400 on eBay (you can find me on eBay at Angelsplace2012). Not bad for a rookie!

That was nine years ago, and, like a growing number of Americans, I've been addicted ever since.

You might think dumpster diving is for people who are homeless, poor, troubled, and so on. I'm here to tell you that it's for everybody. I like to call diving a form of modern-day thrifting. Not only is it economical but it helps the environment by sending less stuff to the landfill. Dumpster diving has helped furnish our home—and given me a wardrobe more fabulous than I could ever have dreamed of! It has created a new income stream for my family through online auctions and has allowed me to provide for those in need through charity. This is an eco-friendly way of living that embodies the spirit of "waste not, want not" that our parents all taught us. I'm determined to make it the norm. And I think more and more people will join the movement when they realize how lucrative it can be. In 2019 I boosted our income by at least $10,000, and 2020 was even better.

In my videos I take my followers on my diving adventures. As I tell them every time, today is going to be an ah-mazing day: "Hopefully we can find something great, but if not, as always, we've been blessed, and we've been blessed with a lot more."

If I have questions regarding an item or situation, I'll ask them for advice or feedback. I share my finds and discuss what to keep, what to sell, and what I'll be donating to charity. I also share my personal life, the good days and the not so good. My husband and kids (usually!) enjoy participating in my videos. We're members of The Israel of God church, where my husband, Antwan, volunteers as a teacher. As a person of faith, my larger purpose is to share the belief that we are all part of something bigger than ourselves, and to emphasize the power of giving back. And so our mission statement is quite simple: Faith. Frugal. Fun & Family.

Diving has become my passion, transforming my life in ways I never imagined and helping me see the world in a new light. I'm grateful and humbled to share this passion with you.

———

As described on Angel's YouTube channel:

Antwan	The man of the house
Taylor, 19	Our camera-shy, funny, and caring teenager
Josiah, 18	Our laid-back, helpful, and charismatic teenager
Hailey, 9	Our princess who has everyone wrapped around her little finger
Samuel	Born in December 2019! The new addition to our family

This is my family. We took this photo on September 6, 2020,
dressed in denim and white for our first professional photo
with Samuel, the new addition to our family.

The more I dumpster-dive, the more it makes me think about how much some people have, while others are in terrible need. I enjoy diving for the thrill of finding things, but I also love it because, in a small way, it allows me to spread the wealth.

I get such a rush from giving. That's one of the things I try to highlight in my videos.

Diving in the Time of COVID

As I was writing this book, COVID-19 was continuing to spread across the country and the world. Regardless of what the case numbers are in your area, you should remain vigilant in taking steps to protect your health. The following is taken directly from the CDC website: "Based on data from lab studies on the coronavirus and what we know about similar respiratory diseases, it may be possible that a person can get COVID-19 by touching a surface that has the virus on it and then touching their own mouth, nose, or possibly eyes. This isn't thought to be the main way the virus spreads, but caution should always be taken." (I advise caution at all times when dumpster diving, pandemic or not.)

If you are interested in diving, do as I do and always wear protective gear to keep yourself safe (I discuss this in Chapter 3) and bring along wipes and disinfectant.

I don't know if it's possible to call it a silver lining, but it took a pandemic to provide the opportunity for many Americans to hit pause and examine their spending habits. The economic slowdown has forced many of us to focus on basic needs and shifted our priorities from the tangible (the latest iPhone, a new sofa) to the intangible (our health, our families, our spirituality). It has underscored the long-simmering issue of income inequality and, I hope, has encouraged more people to see the benefits of acting for the common good. Because everyone is juggling all aspects of life—family, work, school—while spending more time at home, most of us have been rethinking our relationship to our things and our living spaces. Small businesses—a few of which I highlight in the book—have also been forced to pivot in order to accommodate our altered lifestyles. Some of them have even started in response to our new needs and priorities.

If anything, COVID-19 has sort of legitimized the "work" of dumpster diving. I now consider it my full-time job, and whenever people question what I do for a living, I tell them it's all about a larger mission: The things I gather, whether to keep, donate, or resell, are high-quality, perfectly good items that were left on the street and otherwise headed to a landfill—the ultimate in wastefulness. Through diving, I've become a firsthand witness to the excesses of consumerism, and I often marvel at how casually people cycle through the latest trends in fashion, furniture, and electronics. Almost every day I come face-to-face with our throwaway culture, feeling like I'm on a mission to rescue useful, desirable, and life-improving products that other folks have chucked. Again, think about it in terms of ecology and conservation: You can buy the most environmentally conscious product in the world, but that would still be more wasteful than reusing an existing item that has years of life left in it.

It's true that we Americans do love to buy things. In fact, it practically feels patriotic to pile up the consumer goodies. (Consumer spending happens to constitute 70 percent of this country's gross domestic product.) But that doesn't always mean we can afford them. Before the pandemic struck, most Americans were already living beyond their means. According to a 2017 study, nearly 80 percent of American workers were living paycheck to paycheck—and not only minimum-wage workers but those earning $100,000 or more.[1] No matter where you are on the economic ladder, you're probably overspending. It turns out that fewer than half of all Americans have enough savings to cover an emergency expense of $1,000. That's a slim margin. And yet, in spite of this, we feel compelled to spend and spend—on gadgets, on travel, on this season's latest "must-have" items.

The past year or so has seen a lot of Americans breaking out of this mode, reevaluating their attitudes toward money and material objects, brought back to that time-honored value of frugality by sheer necessity. When this virus runs its course, will we simply resume our materialistic ways? Quite possibly, yes. But I hope there will be some lasting takeaways from this once-in-a-lifetime experience, as terrible as it has been. I hope we can all keep thinking sensibly about long-term, sustainable purchases. I hope we can remember to practice humility. And I hope we can remember to be kind to one another.

Inspiration: Community Fridges

One of my favorite silver linings of the COVID-19 crisis is how it's made people more aware of those in need, especially their neighbors. There's no better example of this than community fridges. You may have seen one appear in your own neighborhood recently. They're usually colorfully painted, and people stock them with fresh food and drinks. Everyone is welcome to take what they need and encouraged to bring anything extra they might have. Volunteers keep them stocked and clean, and stores and restaurants sometimes add their excess food at the end of the day. Some fridges in places like New York City have expanded to include toys, clothing, and other basic family needs. The response from the community at large has been overwhelming—often the fridges will be empty less than an hour after they're stocked, which just shows how many people are hungry right now.

It's worth going online to see if you have a community fridge in your city or town. Sometimes the folks who operate them will upload maps with their exact locations. And if you do have some time on your hands and want to help, they're always looking for volunteers. It takes a village to keep the fridge full, running, and safe from vandals.

1

How Our Faith
Drives Our Diving

One important aspect of my diving involves my faith—my name is Angel, after all—and my church, The Israel of God. I've been surrounded by faith literally since my first day on earth, because I was born with ventricular septal defect (VSD), which is a hole between the lower chambers of the heart. I had my first open-heart surgery when I was three months old, and at the time the doctors didn't know if I would make it. But my name turned out to be a premonition, because I had an Angel watching over me. My walk with God started at an early age, and now I'm living testimony of God's amazing healing power!

My faith eventually led me to The Israel of God, which has shaped my adult life in profound ways: I met my husband, Antwan, at church over ten years ago. Israel of God teaches the "uncut" word of God, according to the prophets and apostles. We believe in the words as they were written—that God's will is all there in the text. As a result, we keep feast days and observe the seventh-day Sabbath (from Friday evening to Saturday evening), and we don't eat any unclean food, as defined in Leviticus 11 and Deuteronomy 14 (King James Version). We say that while our faith is Christian, our nationality is Israelite. The Israel of God practices humility and tolerance, and is open to all comers, no matter your race or creed.

Antwan and I each discovered The Israel of God in our own way. I grew up Lutheran, until around age eleven. Then, after my family moved to the South Side of Chicago, my grandmother—who was a big influence on my spiritual life, as she helped my mother raise me—came across an Israel of God pastor while watching TV (its headquarters are in Chicago). She decided to attend a service to learn more about it. She was taken with the church's community and message. After she came home, she announced we were joining The Israel of God. Because I was a child, I had no choice but to go. But by the time I turned eighteen, I was ready to leave the church. In spite of my chronic health

Returning to church after maternity leave was a great moment. I finally
got the opportunity to introduce Samuel to my church family. In the midst
of all the onlookers, I managed to slip away for a quick family photo.

issues—including getting a pacemaker implanted when I was fourteen—I was a typical teenager who wanted to be independent, explore, and have fun. I wanted to see what else was out there.

I met Terrell, my daughter Taylor's father, when I was twenty-one. A few years later we separated, and I was forced to move around a bit, staying with my sister Angela. I was trying to provide a sense of stability for Taylor, but the reality was that I had no control over my situation. During this time, my grandmother continued to try to persuade me to go back to church; she knew I wouldn't be able to move forward without figuring out where God fit in.

At this point, I got very sick with hyperthyroidism. Because I had been so busy working and parenting, I wasn't paying attention to my body. I had lost weight, my eyes were yellow, my skin was dry. I was working as a teacher's assistant, and one day I could barely make it up the stairs in school. I finally went to the hospital, where they discovered my organs were about to shut down.

This medical ordeal was the wake-up call I needed to get my life together. I promised myself that if I got well, I would go and serve the Lord wholeheartedly. I recovered, and the first Saturday after I got out of the hospital, I returned to The Israel of God. Antwan was there reading for the pastor, and I knew at that moment I had to go and meet him. I still remember the name of the lesson: Marriage and

Family Conduct. Who knew this lesson would be so instrumental to our life as a blended family?

Speaking of that, some folks have asked me how the kids feel about being a blended family, since it's an experience many parents will have at some point. I tell them that we had our hiccups at first, of course—Taylor thought Antwan had too many rules!—but now they couldn't be happier to have a "bonus mom" and a "bonus dad." Importantly, though, we don't actually use those terms—nor do we use "step" or "half," although I understand why some families do. I, of course, feel that Josiah, Antwan's son from a previous marriage, is every bit as much my child as my other children, and I know Antwan feels the same way about Taylor. So the example we want to set for them is one of valuing every member of the family equally—no "half" measures!

In every family, blended or not, there will be some give and take. (As Hailey hilariously puts it, having big brothers and sisters can be "annoying.") But because we spend so much time together and really emphasize the idea of ourselves as a unit, our living for one another is second nature. As Taylor now says of Antwan: "He does a lot for me, and gives me great advice!"

Antwan grew up in a Baptist home in Chicago and attended a Catholic school, but from a young age he started questioning religious doctrine. His mother was killed at age twenty-five, when he was just eight years old, which contributed to his skepticism. He read the Bible and saw how the word of God could be interpreted by different religions: He was learning about Catholicism in school, but at home his family followed Baptist teachings, and an uncle took him to Baptist services. As a teenager, he thought he was done with church.

Then, during his sophomore year of college, as he approached his twenties, he began to think more about his mother's death. "I realized how young she was when she was killed and started thinking I wanted to better understand God before I die," he said. That summer, someone introduced him to The Israel of God. It instantly appealed to him. Antwan said, "I had been struggling, and I liked the idea that God has provided answers directly as it was written in the Bible. There are no interpretations; it's not about whether you agree or disagree—it's about studying the text. It's not about anyone being chosen or special." After a few years, Antwan was asked to become a teacher in the church. As he explains, "In The Israel of God, every time you go to church, we call that a lesson. A lesson is about one and a half to two hours of Bible study. Teachers lead these lessons."

So, all of this leads to the day I first saw Antwan teaching at church. My brother-in-law Thein Spencer is also a teacher, and I asked him to make an introduction. The rest, as they say, is history. We were fortunate to be able to blend our families, later adding Hailey and Samuel to the mix.

I also owe my dumpster-diving success to The Israel of God. After we got married, Antwan decided to go to graduate school to get his MBA. I wanted to try to help out in some way and generate extra income for our family. In addition to graduate school, we had just moved to a new house and were looking for inexpensive ways to furnish it. I was trying to find a part-time job using my degree in organization management or something I could do from home, and that's around the time a fellow parishioner mentioned she was planning to "go diving" the next day. I had no idea what she meant, and she rather sheepishly explained that it had to do with searching through other people's garbage and recycling bins. I wasn't so sure what I thought at first, but then she shared some pictures of the things she had found. They were nice items—stuff I had never imagined anyone would just put out with the trash. So I decided to give diving a go.

One of my YouTube followers pointed out that the Bible actually has a term for diving: "gleaning." In fact, that word is used in one of my favorite Bible stories, that of Ruth and Naomi. (Ruth collects the remains and leftovers after the harvest, known as gleaning.) I was moved that someone thought to make that comparison, and find it apt. I believe that my religion, following the Bible, has taught me to live humbly, with intention, and to be kind and compassionate to others. My faith helps keep me positive, teaches me to appreciate the present, and I think these values also help attract my YouTube audience.

When YouTube sent me a Creator Award for passing 100,000 subscribers, I posted:

A woman who knows what she brings to the world isn't afraid to share it! Thank God for giving me a platform to share my gift with so many people around the world! This isn't just an award but a constant reminder of God's light not hiding under a bushel!! I will continue allowing him to use me for his glory PERIOD. Also, I must give thanks to everyone (Subscribers, Family, and Friends) that supports me on this journey . . . your love and kindness can't go unnoticed. Thank you and I love you ALL!! ❤ ❤

And I feel that even more today.

In terms of my social media platform, I try to provide a sense of community—especially now that it's hard for people to gather in person. I use it to reflect our values and how we live as a family of faith. I also emphasize the power of charity and giving back. I know so many communities are really struggling now, and ever since I started diving,

I've kept an eye out for high-need items I can donate to organizations like Goodwill and the Salvation Army (winter coats, water bottles, canned goods, and socks). I'll occasionally get specific requests from friends and neighbors, who know I'll eventually come across what they're looking for (a lamp, a desk, shelving).

In short, I've always known the Lord was with me every step of the way. I'm here for a reason, to change people's perspectives and open their minds to the concept of diving. I truly believe every find is a gift from God.

And the best part is when my followers weigh in to support the mission. Back when I first got started, Kia Serrato said:

> Love watching your channel you give me so much inspiration and I feel so blessed to know that people like you exist always giving love the lessons I learn from watching you and pray that I can be more like you. Amen, God Bless.

And just a few months ago, Jeannette said:

> Just started watching you. I love what you do and are able to reuse or give to someone. Or even make some extra money. I'm all for keeping all the junk out of landfills. You are very nice and very pretty inside and out.

Commenter Gina Evand agrees:

Praise God, That's a wonderful blessing. I agree with you on paying tithes. The Lord comes first. It's the right thing to do. Such an uplifting video. I will also agree in prayer with you that you will find the truck of your dreams!

And Janice Bilbrey, who acknowledges not being religious herself, still sees the value in reusing good things:

Just had to let you know how deeply grateful I am not only to see you repurposing items to help make money for you and your beautiful family but donating to other needy souls. I'd say I'm very spiritual but not a religious person, although if I went to the church you & your family go to I might end up with a little religion in me.

Inspiration:
Bernie's Book Bank

One of the things I always find plenty of while diving are books, and it makes me wish people didn't feel they had to throw them away. I know it can be hard to find a place to donate adult books, but if you have any children's books your kids have outgrown, and they're not totally chewed up, please consider donating them to one of the many organizations committed to getting books into the hands of every child.

One of these near me is Bernie's Book Bank. They've distributed over a million free books just since the start of the pandemic, and twenty million books since their founding in 2009. And they offer plenty of ways to help—by donating gently used books; by volunteering to pack and ticket books or organizing a book drive at your school or community center; or simply by donating money to purchase books. Just $12 will buy twelve quality books for a child who needs them. And Bernie's commits to giving that child another twelve books every year, appropriate for their advancing reading level, until they graduate high school.

All the research tells us that access to books at an early age makes a huge difference in kindergarten readiness and can keep children from falling behind even years later. Yet two out of three low-income children have no books of their own.[1] And because so many schools were closed during the 2020–21 year, they're at risk of losing even more ground.

If you're not local to Chicago, check the internet—there are plenty of book-based nonprofits that operate on the same principles as Bernie's. Or check your local libraries and elementary schools to see when they hold their book drives. (For that matter, library sales are another great way to get quality discount books for yourself and your family.) I know the precious moments I've spent reading books to my babies are some of my favorite mama memories, and I wouldn't want any child to miss out on that.

2

How to Get Started

As I mentioned, it was a fellow church-goer who introduced me to the world of dumpster diving.

On my first dive I went to an upscale suburban neighborhood nearby, and sure enough, I was amazed at the range of mint-condition things left curbside: clothes with the tags still on, unworn shoes in their boxes, unused gift cards, game consoles with no signs of wear and tear—you name it. After that first foray—in which I found the playground set that our then toddler loved—I was hooked.

Initially, I limited my scope to things my family could use. But I started feeling guilty about leaving perfectly usable items sitting out on the street, most likely destined for the landfill, so I began taking additional things I knew could donate. I had also been looking for new ways to get

erate income and realized I could resell some of my finds on eBay. (As one of my followers once put it, "Slay all day and sell it on Ebay!") These three approaches—*use*, *donate*, and *sell*—continue to inform where I go diving, and how frequently. (I used to try to go diving every day, but at a certain point, my family started to feel neglected, so now I venture out two or three times a week.)

In the early-morning hours before I start my day I usually go to my prayer room. (I have shown this room to my followers.) It's just a small closet, with a student desk that I actually found dumpster diving, my scriptures written out and pasted on the walls, and my Bible. I use the quiet time and space to talk to the Lord, or pray, or just meditate on scripture. Then I make myself some tea and I'm good to go!

Ready to try it? Good! But before you go, I would like you to ask yourself this simple question: *What am I diving for?* In other words, what's your purpose? A dive can certainly be aimless fun, a "let's just see what I can find" adventure. After all, most people probably start the way I did, out of simple curiosity. But if you take some time to think of a few specific finds, you won't feel compelled to bring home the first thing you see. You don't have to settle: If you keep at it, you will eventually find the thing you're looking for. Trust me, it's out there somewhere. And if you don't really have extra time to donate or sell, don't bring home things you'll never use or something you think you might

upcycle at some point down the road. Trust me, your storage space will thank you. The best diving is fun but focused.

To get an up-close look at what diving is like for me, subscribe to my YouTube channel, MomtheEbayer101. The reason I film and share my diving adventures is to inspire others to give it a try. I post videos daily and explain my finds. One of the best things about building a community of followers is I never feel alone when I'm diving; I know they have my back. In fact, one of my YouTube followers started a GoFundMe for us, so I could afford a truck and make it easier to haul my finds. The generosity truly floored me. As of November 2020, I became an ecstatic truck owner!

Angel's Top 5 Finds:

1. $1,000 cash (yes, cash!)
2. Two stand-alone glass shelves I'm using in our dining room
3. Refrigerator
4. $600 Neiman Marcus gift card
5. Vintage 1960 JFK campaign poster

Antwan's Top 5 Favorites:

1. Nespresso machine
2. Laptop
3. That $1,000
4. Standing lamp for office
5. Basketball hoop (and then a basketball!)

Our kids' favorite items:

1. Xbox
2. Designer clothes, jewelry, purses
3. Mirrors
4. Desks
5. Arts and crafts materials

Fun fact: When I asked my kids if any of them would consider diving on their own when they're older, they all said "No!" Of course, they're out there with me now, finding their own good stuff. Taylor is picky and very girly, so it's not easy to find stuff for her, but she has jewelry, purses, and branded stuff that she likes that we didn't have to buy. Josiah loves it—he gets watches and high-end executive shirts, and I even found an Xbox for him once. And all the cuddly toys in Hailey's bedroom have come out of the dumpster. I get them home and wash them, and she loves them. And, of course, half of Samuel's toys are dumpster finds!

How to Know Where and When to Go

I've developed what I call a "spidey sense" for finding the best dumpsters. Naturally, it's a process of trial and error—more of an art than a science. But I have some basic principles on lock. Like most things in life, if you invest a little time and effort up front, you'll waste a lot less time down the line.

- **Know your territory.** You'll need to spend time getting to know an area's normal routine—what types of items are being left out on a regular basis, and when. There's a good chance you might already know this, if you're planning on diving in your own neighborhood or town. When I got into diving, I made sure to explore a range of neighborhoods around Chicago and its suburbs. I would go back every other day or so for a couple of weeks to familiarize myself with trash pickup days. I normally stick to upscale residential areas, rather than chain stores and malls, but the same rules apply to retail dumpsters too. Some stores throw out overstock or expired inventory every week; some do it once a month. If you scout a location consistently, you'll soon learn its rhythms.

- **The early bird gets the worm.** I'm typically out the door no later than five thirty in the morning to go diving. It's partially to accommodate my home life and my kids' school schedules, but it's also invaluable to get out there before anyone else comes along to claim things that were put out the night before. Some divers even like to go in the middle of the night—especially those who focus on retail—but as a woman who often goes out on her own, I feel safer in the early-morning hours in terms of visibility and how others might respond if they see me.

- **To everything there is a season.** There are reliable human behaviors around the changing seasons and holidays. These annual rituals and transitions usually promise more treasure. A few key times I'm always sure to hit:
  - Spring cleaning is definitely a thing! Starting around April, there's always an uptick in the number of items left curbside.
  - College campuses and their surrounding housing areas are great to explore when school lets out. Students (and their parents) are usually in a rush to pack up and leave. I have found unused gift cards worth hundreds of dollars still tacked to bulletin boards; barely used computers and phones; game consoles; and clothes with tags still on them.
  - People tend to get rid of stuff before the major holidays, either because they're

preparing to host guests or bracing for a flood of gifts. The new year is also an auspicious time, when people throw out those gifts they didn't need or remove older items to make room for the new. Last year I found new parts for a drone still in the packaging, plus loads of books, puzzles, Legos, a basketball, a new backpack, and a bunch of beautiful clothes. I once found a brand-new sweater still in the Dillard's box! So just a reminder, folks, that if you got some stuff this year that you're not crazy about, don't throw it away—there are so many families feeling the squeeze! Take it to a donation center or post on a local Facebook group for exchange.

I love when my followers post about even more amazing post-holiday finds! One recently wrote:

A friend was dumpster diving in Chicago Alleys the day after Christmas and found a gently used Hermès Birkin bag with $2,300.00 in it. The Birkin bag has been authenticated as real. What a find!

Can you believe that?! When I responded to her in the comments section, she told me she tried to find the owner of the bag, thinking it must have been thrown away by mistake, but she wasn't able to. So she kept the bag and donated some of the money to charity—paying her good fortune forward! This is a great example of how we can embrace blessings for ourselves without forgetting to give something back in gratitude. When it's not spring or the holidays, keep these additional life rhythms in mind:

- The end of each month is a good time to get out and explore, because that's when people typically move.

- Home renovations and homes for sale: If there are signs of a home renovation under-way, then I know to keep circling back—contractors and decorators always throw out valuable home-improvement products, such as cans of unused paint, leftover wall-paper, and boxes of ceramic tile, along with the old furniture, knickknacks, and house-wares they plan to replace. I've picked up spare cans of paint from premium mak-ers this way, happily using the product to freshen up bookshelves, dressers, and chairs in my house.

- Realtors will occasionally put out quality items after staging a home, which can mean a bonanza for a diver: patio furniture, firepits, home decor, picture frames, side tables, and the like.

- Estate sales: If I happen upon an estate sale, I'll stop to see what's available. The organizers usually have many leftover or unsold items they're willing to give away. I've "inherited" some beautiful vintage things. If they're available, I'll also take unused personal hygiene products to donate to homeless shelters.

Retail Stores

Although I generally prefer to stick to residential diving, there are divers who focus on retail dumpsters. Stores cycle through inventory on a regular basis, and will throw out old stock, new stock in damaged packaging, and returns.

These stores are particularly popular with retail divers:

- World Market, Dollar Stores (home decor and trinkets)

- PetSmart ("expired" pet food and toy over-stock)

- Old Navy (clothes)

- Best Buy, Office Max, and GameStop (discontinued or returned home electronics and gaming supplies, returned office furniture, photo paper and toner, CDs)

- Ulta Beauty and Bath & Body Works (make-up, cosmetics, skincare)

A word of warning: Be alert around retail dumpsters. Stores may deploy security guards to check what you're doing. Also, some big-box stores use commercial trash compactors (which generally look like a dumpster attached to a larger enclosed bin)—these can be dangerous to open and rescue finds from.

Another interesting strategy I've heard of to find used

items (but haven't done myself) is to rent out a small unit in a storage facility in order to gain access to the things people leave behind. When people move their stuff out of storage they have to make decisions about what to take with them. Inevitably they leave things behind, either in or near the facility's dumpsters. People who move, divorce, remarry, or are coming to collect the possessions of a deceased loved one often toss an amazing array of valuable items. Just by visiting the facility you could stumble on some major finds. (An added bonus: They're right out in the open!)

A few more general tips, once you get in there:

- Boxes sealed with tape can indicate that good things are inside. Don't ignore!

- You'll surely find discarded suitcases, duffel bags, purses, garment bags or trunks, and dresser drawers. It's all great stuff, but make sure to look inside! A discarded Tumi carry-on is a fantastic find in itself, but the serious treasure might be sitting in a zippered compartment.

Finding Your Treasure

Things I'm always on the lookout for:

- Furniture
    - chairs
    - dining tables
    - side tables
    - entertainment centers
    - standing shelves
- Baby gear
    - baby clothes
    - booster seats
    - rockers
    - cribs
    - toddler bed frames
- Books/DVDs
- Dishware/glassware/silverware
- Picture frames
- Candles/votives
- Electronics
    - old laptops
    - phones
    - game consoles*
- Gift cards (I never, ever turn my back on gift cards—always take gift cards. It's easy to check the remaining balance!)

- Luggage
- Lamps
- Appliances
  - blenders
  - toaster ovens
  - mixers
  - SodaStreams*
- Rugs
- Games
- Toys
- Clothes (from designer labels* to basics)
- Shoes*
- Old black-and-white family photographs*

* These items sell best on eBay.

My Dumpster Designer Wardrobe

So let's talk about the clothes! Everyone who knows me knows I'm a sucker for fashion. But what they may not know is how much of it comes straight from the "trash" or thrift stores like Goodwill. I've found Burberry scarves, Armani pants, Pandora jewelry, you name it! And while it feels great to save money, part of the fun is the hunt, and how pieces you might not have picked up in a store can

Some of my most memorable finds! An outdated coat that I took from Goodwill and spruced up (*top*) and a Ralph Lauren dress I found at Goodwill with the tags still attached (*left*); and my celebratory photo shoot after reaching 10,000 followers (*right*).

make your whole look pop. I show off some of these looks on my Instagram page, @momtheebayer101, and reveal where the pieces came from. I also get great feedback and style inspiration from my sisters in the eco-fashion community.

For instance, on my last Goodwill trip, I found a killer leather duster jacket and a pair of printed pants. I spent just $30 for a whole haul of stuff I thought would *perfectly* complement my wallet, shape, and style. Although I stepped completely out of my comfort zone with the pants, which I normally wouldn't wear, I'm so glad I gave them a try.

You see, style is more than looks—it's attitude too. When you throw a little fierceness into your wardrobe, honey, you can make a paper bag look like top-model fashion!

And sometimes there's no need to hold out for designer when you can make the inexpensive stuff look just as good! On January 29, I posted about how I re-created a $275 purse the eco-fashionable way!

If you're diving or thrifting in the hopes of spicing up your wardrobe, remember that there are style ideas all around you. Look at what other folks are rocking and figure out ways to make them your own. Don't worry about copying—if someone else has laid the groundwork and it still works, there's no need to reinvent the wheel!

Download Pinterest, go into your closet, and see what

Finding Your Treasure

momtheebayer101   ...

       • •

momtheebayer101   "Hi Ladies!! We all love a cute bag/ purse, but sometimes our pockets simply can't afford it. Well, I figured out a way to have the high end look for more than half the price.

Wait.... before I move forward—please don't claim the purse you create to be the actual purse you're inspired by, because that's illegal! I'm not bailing anybody out of jail!

Now! I sourced my local Goodwill and found a similar bag. Next, I went to @forever21 and caught a MAJOR jewelry sale (Buy 3 get one free); I bought the necklace for $7.99. Finally, I attached my necklace to the purse and VOILA!!! Remember, you can get fashion inspiration from everyone and anything, but your style will always be one of a kind!"

pieces you already have; now you know what to look for when you're out on the hunt!

And one final note: If someone gives you a compliment, embrace it! Devaluing yourself doesn't equal humility. When someone recognizes your beauty, accomplishments, or anything else that makes you great, don't highlight your flaws!

Graciously accept those compliment(s) with a simple "Thank you." Remember, you are worth it!

Bringing Back the Pack!

I really didn't care too much for fanny packs until I found this crossbody purse in the trash and turned it into a fanny pack. (Yup! Someone threw it away—in the garbage.)

My outlook about this old-school trend changed instantly after finding one for FREE! Now that I know there's a modern, chic, and stylish alternative to the unattractive versions I saw in the eighties, I'm not opposed to wearing them anymore. In fact, I'd prefer a chic fanny pack over a purse anytime because of the ease of use. Like the old sayings go, some things just get better with time, and one person's trash is definitely another person's treasure!

I love to use the main street close to my house as a runway.
Here's me modeling a white crossbody purse I converted into
a fanny pack. I'm also wearing a fur-hooded coat liner
I scored from a dumpster to complete the look.

Inspiration:
Dress for Success

Lots of new charities have popped up in response to the pandemic, but I'd like to give a shout-out to one that's been around a long time: Dress for Success. The premise is simple—to help women get financially independent by providing them with professional clothing and career support. This is a cause close to my heart as a mother, and I know how many ladies are out there trying to work and support a family at the same time. It's not easy to keep up, especially with so many jobs going part-time or temporary.

Dress for Success operates boutiques in almost 150 cities around the world. There, volunteers help women choose outfits in advance of job interviews and offer coaching on interview strategies as well. But it doesn't stop there. The organization offers support through the job search and beyond, and many of the women it serves end up volunteering their own time, donating clothes, or starting similar initiatives within their own communities.

I think this is a great example of how thrifting and consignment isn't just a hobby or a way to look great for less—it can be part of a mindset that changes people's lives and creates a cycle of giving we could all use more of!

3

How to Be Prepared (and Stay Safe!)

As a woman going diving on my own, safety is always my priority—for both my person and my health. Here are the safety tips I've developed over the years.

Wear old clothes and sturdy gloves. You can buy needle-resistant gloves at hardware stores, garden centers, and cookware shops, intended for a variety of applications. They're also available on Amazon. Personally, I wear disposable plastic gloves under sturdy protective gloves. The protective gloves can be a little clunky, so if I drop something or see something small I'd like to pick up, I can take off the thicker gloves and more easily grasp it. Comfortable shoes are a must. I wear a Frogg Toggs protective suit and a

I spend a few minutes after filming each YouTube video trying to capture the perfect shot for an eye-catching thumbnail. After all, thumbnails grab the audience's attention before the videos do.

safety vest with reflector lights for extra visibility (both are available on Amazon). As a profile on me in the British newspaper the *Mirror* noted, "Angel Williams takes her 'job' so seriously, she spends around $50 (£40) a month on protective clothing so she can jump into rubbish skips and retrieve precious items." (I love how the Brits make it sound even more fabulous!)

I always recommend—and did even before the pandemic—that anyone diving should wear a mask and/or plastic face shield. Bring paper towels, a rag, or wipes to clean your hands. Hand sanitizer is a must, along with sanitizing wipes to disinfect your car, your gloves, your steering wheel, and your door handles. You'll also want to wipe down some finds before putting them in your car.

Bring along extra bags, a blade for slicing bags and boxes, a step stool, and a trash picker or grabber tool for reaching into tall dumpsters, pushing trash aside, or poking holes in bags to see if there's anything of interest inside. If you night-dive, bring a flashlight or headlamp.

Another one of my favorite items is a fanny pack. It holds a few things that I want within easy reach, and I can deposit smaller finds in there. Some things I carry in my fanny pack include hand and toe warmers (Chicago gets brutally cold in the winter), a marker, tissues, wipes, and my business cards. I also carry change in case I encounter panhandlers.

Don't listen to music or wear headphones—be aware of your surroundings at all times. To feel extra secure, you may want to bring pepper spray. Also, keep in mind that there are times when folks are feeling more nervous than usual so they may approach or question you even if they normally wouldn't. This happened to me most recently in August 2020, when the Black Lives Matter protests were in full swing in Chicago (and across the country). For a while it was hard to even get to my usual places (or home again afterward) because some streets were blocked off or closed. One man approached me, asking who I worked for and why I was filming. When I told him that I worked for myself and was only taping myself, he looked a little confused, but he kept on walking. I try to always have a big smile and be up-front with folks about what I'm doing, and it never hurts to take that advice even more to heart in times of extra stress and conflict.

On a related note, if you see a maintenance or sanitation worker approaching and you think they may not see you, it's always a good idea to just pipe up and make yourself known. After all, they need to empty and sometimes lift the cans, and you don't want to get hurt! Once a custodian even offered to get me some bags that were out of my reach. People are mostly helpful and kind.

Even though it's called dumpster diving, I would never

actually dive headfirst into one! My number one rule is to be very cautious when going inside a dumpster. There can be heavy articles and broken glass that can cause injury, gooey messes you don't want to step in, and the occasional critter searching for food, who won't be very pleased to find you poking around their territory. You can usually do all the diving you need while safely outside the dumpster. In order to see better and to reach inside, I bring along my grabber and a folding step stool.

I try to take only what's outside, next to the dumpster, or what's inside on top of the pile. In some cases, you can plumb what's easily visible and just a few inches down. I do not advise digging deeper than that, as that's where rodents usually lurk.

If the stuff in the dumpster is piled mostly to one side, work your way from the low side to the high side. If you start on the high side, items will tumble and you'll have a hard time keeping track of what you've seen.

Don't take what you don't need, and leave the dumpster area as tidy as you found it. I know, it might sound ironic, but be respectful of your neighbors and the folks who collect the garbage.

Once, I found boxes scattered around a retail dumpster, and it was obvious someone had been there diving before me. Here's what I had to say about that:

Let me show you, look at this. They just left all of this stuff out here, on the street. Attention, dumpster divers: Listen, this is unacceptable. Look at this mess! If you're going to be out here in the world of dumpsters, you've gotta abide by the Dumpster Diving Rules. What are the rules? Clean up after yourself. What are the rules? Be kind to people out here. What are the rules? Share. What are the rules? Take only what's necessary. What are the rules? Be courteous. What are the rules? Enjoy the adventure. When people see this type of stuff, you know what they do? They automatically want to ban dumpster divers. Why? Because this is disgusting. Now I have to sit out here and clean this mess up. I pray that all of us are able to find some amazing stuff, I pray that we all do well out here. I'm not trying to be mean, but I just wanna warn you guys. If you do not clean up after yourselves, they are gonna shut us down!

Helpful Hints
from the Community

My YouTube followers have been so helpful in suggesting items for divers to keep in their pockets or vehicles while on the hunt. These include Band-Aids, extra garbage (or tote) bags, a whistle, and—this one is my favorite—a battery-powered outlet to test electronics on the spot! (A subscriber gave me one—thank you, subby!) And it took me a few years, but after one of my followers suggested it, I start putting the rear seats of my car down to accommodate furniture. (Yes, I needed someone to tell me that!) Commenter Angie Blackmon says:

Glad to see you out dumpster diving. And a rule of thumb is try to leave it cleaner than you found it, but some don't care & it makes it hard on us dumpster divers. Angel I have a bag that I keep all my dumpster diving stuff in like gloves, headlights because I night dive, sanitizing wipes, hand sanitizer, scissors, knife etc. & a bucket to stand on, rake, grabbers, extra bags for when I find those little things I grab one of my bags to put stuff in & containers so nothing won't spill out into my car, I load everything up before I go so I will have everything that I could think of that I will need. Keep up the good work and keep on diving, be safe & God Bless you!

And KC Fashion suggested:

Angel, keep a box cutter or pair of scissors in your glove compartment for your packages and other things!

Bringing It All Back Home

So, how do you get your treasures from A to B? Put an old towel, blanket, or tarp in the cargo area of your car to protect it from any dirt or debris from your haul. A good cart is also essential for lugging bigger items back and forth.

Dedicate a space in your house for your finds. I use my basement. Create an organized storage area: This can be as simple as a space that can hold a small collection of cardboard boxes. I'll divvy up my finds into various zones (clothes, shoes, accessories, appliances), using the boxes for sorting. If I've collected clothes to donate, I run them through the washer and dryer before packing them up. For eBay, I stick mainly to electronics and quality designer clothes, so I'll clean those items, photograph them, and store them on shelves or a clothing rack until they sell.

In fact, Antwan recently suggested that it's time to get rid of some stuff from my own collection, but I'm not sure what I want to part with. In a happy coincidence, one of my Instagram followers asked whether I'd be selling the looks I post. I always hate to say goodbye to my favorite pieces, but it's good karma to keep the fashion circulating! And as a top-rated seller for nine years on eBay, I definitely have what it takes to get started. So stay tuned if you're interested in shopping my closet!

Is Dumpster Diving Legal?

Good question. Mostly yes, sometimes no. The laws around diving differ from place to place, and they can change from time to time. It's a good idea to check the .gov site of your municipality for the latest local ordinances pertaining to diving. It comes down to common sense: Always make sure you're diving in a place where it is allowed, do not dive where it's illegal or prohibited (watch for NO TRESPASSING or PRIVATE PROPERTY signs), and play by whatever rules your locality has. Here are a few additional suggestions:

- Make sure your license, registration, and identification are up-to-date.

- If you are approached by law enforcement, community watch, or a homeowner, kill them with kindness. I've had some people approach me out of concern that I'm trying to steal someone's identity, or people who just want to make sure I'm not making a mess or endangering myself. I always greet people with a smile and cheerfully explain what I'm doing. Which leads to my next point: Be honest about what you're doing.

Nine times out of ten, people (including the police) are just curious, not suspicious or hostile.

- If you're going to be diving on a regular basis, I recommend getting business cards printed, as I did. It's easy to do, and it shows the world you're serious!

Inspiration:
The FLOW Foundation

Another heartbreaking outcome of the pandemic is that many women in abusive relationships are feeling more trapped than ever. Calls to hotlines are way up, and shelters are at capacity almost everywhere.[1] So I want to highlight a local Chicago organization that's doing the Lord's work for women in this situation: Kisha Barrett's FLOW Foundation.

When her sister was killed by an ex-boyfriend, Kisha was distraught. She vowed to do something to save others from the same fate. Now, with FLOW, she helps secure housing for women in danger, but that's just the beginning. As the *American Reporter* writes, in a piece from October 5, 2020, "The service acts of The FLOW Foundation do not end at providing resources, shelter, and legal assistance for disadvantaged women; the foundation is taking a holistic and spiritual approach. The main goal is to create a new peaceful and comfortable atmosphere for women so they can grow and get stronger from within. The foundation organizes yoga and meditation classes, confidence-building and self-esteem workshops, self-defense training, and entrepreneurship classes. The goal is for women to get better than they ever were so they can be mentally and physically prepared to walk out of an abusive relationship." And all services are free. In the next five years, Kisha hopes to open several more locations around Illinois and across the US, because she believes the need is so great.

As any abuse survivor knows, recovery is a lifelong process. Without a strong structure keeping you safe, it's a daily battle, especially if you have children to think about too.

Organizations like the FLOW Foundation are helping to stitch together the holes in the safety net so nobody falls through. It's an effort we all need to support, however we can.

4

How to Find a
Diving Community—
Online and Off

G od is the reason for it all, and I'm truly grateful for my online family and their support. Over the course of nine years as an influencer, I've gained more than just "followers"; instead, I gained a community, connections, personal interactions, and loyalty from so many beautiful people all over the world.

My work writing and creating meaningful content has impacted lives in many different ways. I'm forever grateful! And this is just the beginning. I have so much more I want to do.

With the help of the Lord all that my heart desires will come to pass. In the meantime, I'll continue doing my part while remaining humble. If you want to join the community there are plenty of places to start.

Information and Resources on Social Media

When I'm looking to connect with other divers, trying to find organizations that accept donated goods, or just needing inspiration, my first stop is often the internet. It's probably the easiest way to develop a support network, gather important local information, and get fresh ideas.

YouTube

If you're interested in seeing what other divers are up to, YouTube is the place to go. In addition to yours truly, you'll be able to "meet" a variety of people, individuals and couples, from all over the country who are committed to diving. Some divers focus on finding specific retail items they can sell, some concentrate on rescuing perfectly good produce and other food items from supermarket dumpsters, some like to upcycle old furniture, and some try to live sustainably in every way.

For instance, Steve and Steph, a YouTube couple also known as the Resale Killers, focus on diving behind big corporate stores, and the amounts of unused balances on gift cards they find is amazing (not to mention the unused inventory). That Shabby Guy posts amazing restorations of dressers, armoires, tables, and more, and shares all the details, including materials and project times, so his fans can do the same. And the supermarket divers could write a whole book of their own!

There's a whole world out there, and no matter the approach, I think all divers believe in spreading the word that there are amazing treasures to be found in trash and drawing attention to our rampant throwaway culture.

How I Grew My YouTube Following and Created My Own Community

I didn't set out to become a YouTube content creator, but when I first thought it might be fun to share my diving adventures it was mostly because I enjoyed watching other people on YouTube. It was the main platform I was familiar with, I wanted to be able to post videos of varying length, and the community seemed to be based there already. After I started sharing my dumpster-diving finds, I noticed people were responding in positive ways. I began to add personal touches: I created opening and closing credits so that my videos always start and end the same way, and eventually taught myself how to use Final Cut Pro so my editing looks more professional. I know my editing work has paid off when I get comments like the one below, from DJ:

> I love that your format is predictable!!! There's so much stimulation in the variation of finds that having a stable format is 👍 . . . love the adventure!!

I try to make my videos fun and engaging. I love to sing, so sometimes I'll share a little song. I also have some catchphrases I like to say, like "I can take nothing and turn it into something" (i.e., make money from selling free stuff), and, when trying to decide if I should bring something home, "Is the juice worth the squeeze?"

I like to elicit some kind of response or feedback, so I really focus on making my dives intriguing for the audience. My video titles are pure clickbait, like "OMG!! Why Would Anyone Throw That Away??" from December 2017 and "I Hit the Motherload at the Stores this Time" from July 2020. And I end every video with a little dance, just to send my followers off to the rest of their day with a spring in their step. As I posted one sunny day when I was feeling myself in my thrift-store finds:

Finding Your Treasure

Justin Timberlake says it best: Just dance, dance, dance! When you dance, your body releases endorphins. This is a chemical that triggers positive energy and good vibes! It helps improve our emotional state and reduce our perception of pain. So basically, dancing is your cure for happiness!

I LOVE dancing! Whenever I'm presented with the opportunity to do so I'm going for it!! Why? It makes me feel HAPPY & good inside.

Sisters/Brothers forget that you may not have rhythm, a dance partner, you had a bad experience, or you don't like it (start)! Drop the excuses, turn on your favorite jam, get on your feet and DANCE!!

You'll thank me for it later!

It's all worth it when I get responses like the one below, from Jessica Huerta:

That dance at the end can never get enough! Love your videos, kills my anxiety away thank you.

And Jody Stone:

I don't normally leave comments on YouTube channels. But I have to tell you how much joy watching you brings me. You are such a kind and caring person. I am going through a rough time right now and when I need a smile I watch one of your posts. God bless you. PS your family is precious and adorable.

I ask questions about items I find (especially if I don't know what they are) or throw out topics like: "Would you knowingly accept a gift that was found in a dumpster?" I talk about things that might be happening at home. Depending on my mood, I will share pretty much anything I happen to be thinking about. I usually stay away from current events, but I have recently encouraged viewers to vote, wear a mask, and try to have empathy and compassion for one another. Love is an action! Given everything that's been going on, I feel responsible for using my platform to at least try and turn down the temperature and turn up the kindness.

And my followers reflect that sense of kindness back to me. Take this lovely comment from Denise Walberg, for instance, on a recent video where I was unboxing a beautiful bunch of pieces sent to me by a viewer:

> I love how much you show your appreciation when you receive gifts from your viewers. You have a heart of gold Angel, that's why we all love you. You're gonna rock all those beautiful clothes & jewelry too! Someone mentioned a fashion channel. Would definitely love to see you follow through with that. At least think about it. BTW, the pearl necklace you received is meant to be worn with the brooch-like piece up the side so it will lay flat. Have fun trying all your new clothes on!

Or June McKinney, a great-grandma who watches TikTok clips:

> I watch all your videos on my smart television. I have to get on my iPad to like your videos. I just started following you on Instagram and TikTok. I'm a sixty-six-year-old mother of three girls, eight granddaughters, one grandson, nine great-grandsons and four great-granddaughters. I have a full, blessed life. I love your family. You make my day. God Bless.

Comments like that give me life!

After I post a video I make sure to go through and respond to each and every comment. I know people appreciate when I acknowledge them, even if it's just by putting
a heart emoji by their post. I want them to know there's a
person behind the screen, someone who is taking time to listen to what they think. I'm not doing anything elaborate or
fancy—I'm just showing I care. Likewise, I try to say hello
to all the folks who show up in my live chat, especially the
moderators who help me monitor the conversation. When
they're taking the time out of their day to watch the video
and say nice things, it's the least I can do to acknowledge
them. I treat my online family as real family; I have genuine
concern for their well-being. And I find it makes the whole
community feel tighter, as well as welcoming to newcomers.

Facebook

Facebook is another great way to connect with other divers. There are diving groups around the world, public and
private, just waiting for you to join. Just search "dumpster diving" and see what comes up. Local groups might be
more helpful in terms of practical advice for your town or
city, but good tips can come from anywhere!

Pinterest

Many divers like to post their finds along with before-and-
after photos. If you're looking for ideas to fix up furniture

or upcycle clothes or fabrics, you'll find no end of inspiration here. I saw one designer the other day who had salvaged a dresser frame without the drawers and turned it into a beautiful storage piece. It was a great reminder that things don't always have to serve their original purpose to be useful—and worth saving!

Listserv, Freecycle, Google Groups

Most neighborhoods employ local online message boards, which act as forums for people to post questions about anything and everything to do with the community. They often include "curb alerts" for goods being left outside, posts about the best local places to donate, and annual events like toy and clothing drives. Nextdoor.com is one popular site, but if you can't figure out where your forum is, ask a local mom with young kids—I guarantee she'll know!

Information, Community, and Resources in Your Area

Churches and Other Religious Institutions

Churches usually partner with charities, and some charities grow out of the church (the most famous being the Salvation Army). Personally, I volunteer with Margaret's

Village Shelter and Pacific Garden Mission in Chicago, as well as for my church's annual neighborhood giveaways and charitable events. Not only does my volunteer work give my diving purpose but it also provides a realistic sense of what people actually need. I'll often grab something I see on a dive with someone I've met at the shelter in mind—a baby bouncer for a new mom, for example, or a book series for a ten-year-old boy. I've realized that people living in shelters may not have the means to buy basic toiletries or medications, and these kinds of products often aren't carried at Goodwill, so if I find unopened or unused items like toothpaste, deodorant, or shampoo, I bring them to the shelter.

I've talked about how volunteering at the Pacific Garden Mission makes me feel in a couple of videos:

> I had such an amazing time being able to serve my brothers and sisters. They had me on food duty. I had a few moments that had me tearing up, just to see everybody . . . If you've never done it before, you have to do it. It really makes you appreciate the things you do have. So many people are out here struggling, wishing they were in your shoes. Next time I'm definitely bringing my kids. They definitely need this experience. I never want my kids to feel that they are better than anybody, or to mistreat anyone just because someone happens to be having a bad mo-

ment in life. I always want my kids to respect all of our brothers and sisters. Just because you have something in one moment don't mean that the Lord can't take it away from you with the snap of a finger. That's why I'm so, so humble—one minute you're on top, the next minute you could be down on the ground. You just never know. One day I could be in here trying to get food for free, trying to feed my children. I would want someone on the other end serving me to treat me with love, decency, and respect.

Then I asked my followers to tell me what they had done for someone lately that had put a smile on their faces. Here are some of their responses:

Ingrita G wrote, "When I worked at Starbucks, I was told that all the food at the end of the day was either to be thrown in the garbage or taken home. No one I worked with wanted to take anything home, so I took at least two full-sized garbage bags filled with completely edible and fine food. I would hand out most of it to the cleaning crew/ security in my building or pass it on to people who actually needed a meal. The amount of food/products that companies and people throw away sickens and angers me, so thank you for going out there and collecting items that otherwise would sit in a landfill!"

This story hit me right in the heart, because I've seen first-hand how much perfectly good food gets thrown away every day, due to food safety policies and laws about expiration dates. Thankfully, some stores and restaurants, like Trader Joe's, are starting to figure out ways to donate the food without running afoul of the law. Even Starbucks has launched a partnership with Feeding America to donate their surplus food. According to a statement on their site in 2016, "Starbucks intends to scale this program over the next five years and rescue 100 percent of its food available for donation from participating company-operated U.S. stores."

And it works on a smaller scale too. As follower Suzette Fam Espinosa Pina wrote, "I bought breakfast for an old, retired man who didn't have money."

And Paula Smith wrote:

I was at my local Dunkin' Donuts and brought a hot chocolate and donut for a young woman panhandling outside. She just kept thanking me, and I kept telling her no problem.

I love these stories too, because not only are you giving someone a meal they need, you're making them feel seen and cared for. Donating to big organizations that help lots of folks at once is always great, of course, but it's these one-on-one moments that remind us of our shared humanity and make your town or city a better place to live. As Paula found out, most people are so grateful for a gesture that costs so little.

There are plenty of other ways to be generous too! As another commenter shared a few years ago:

> 3 times a month I do crafting with neighborhood children free of charge . . . sometimes all it takes is someone to care . . . I love it.

This reminded me that it's not always about immediate needs like food, clothing, and shelter. Our kids also need adults to teach them, help them use their imaginations, and create something beautiful to put back into the world, especially now, when so many of their parents are working so hard just to make ends meet. We all have the talents God gave us, and it takes them all to make the world go round! When you share your talent and your time, it doesn't cost you anything, but it can give someone else just what they need.

Even my fans who haven't yet jumped on the dumpster-diving bandwagon are moved by the sight of so much going to waste. As commenter Courtney Bush shared three years ago:

> This is my first time watching. Honestly, when I started watching I cringed about going thru trash, then your speech had me in tears. I've never been in that situation, thank GOD, but it made me realize that the stuff you were pulling out shows how much we waste and YES, those

things are valuable when you are down on your luck. I still can't find myself doing it LOL but thanks for what you do!!

I also shared some thoughts with my followers after working in the shelter's women and children's distribution center, where I often bring clothing donations:

I'm just coming on here just to talk about today and my experience, and how I was able to touch so many lives today. I try to volunteer a lot of my time at the homeless shelter. When I first started, I worked in the kitchen. And I got close to the program manager and asked if I could volunteer in the women and children's distribution center. When I got down there, I immediately connected with a sister—oh my gosh, she was telling me her story and how she got into the shelter, and how she turned her life all the way around. Hearing her story and how she overcame addiction and adversity, it was a humbling experience. It made me realize, if you just wake up in the morning, that is a blessing. If all of your members in your body are working and functioning properly, you are blessed. If you have all of the toes on your feet, you are blessed. You just do not realize how blessed you are.

Being at the homeless shelter, you see so many people, from so many walks of life. When you hear their stories, they just keep you to where you are. It's just a humbling state of mind. I'm supposed to be at the shelter, touching lives, allowing the light of the Lord to shine through to those people. They've been down and out for so long, they need someone coming in and telling them how great they are.

Follower Nicola Locke agreed:

I think working at these places really makes you realize that we're all just a few steps away from the same situation. Race, gender, and class have nothing to do with it. Anyone can end up homeless. You inspire me to be a better person. You're helping many more than the people you meet. Thank you!

And Brown Blasian has been there himself:

This is amazing!!! I should start giving back to the less fortunate because I was homeless back in February. But someone like you gave my family a chance! I would like to do the same! You are so inspirational! God bless!

Reuse/Sustainable-Building Organizations

Many communities host nonprofit organizations that specialize in reselling, at huge markdowns, home-building salvage materials, along with gently used furniture and housewares, making reuse accessible to the community. These outlets generally work on donations. Some examples:

HABITAT FOR HUMANITY RESTORES

Habitat ReStores are independently owned reuse stores operated by local branches of Habitat for Humanity, the global nonprofit devoted to the creation of affordable housing. ReStores accept donations and resell home-improvement items to the general public at a fraction of what they might sell for at big-box stores or conventional suppliers. I think Habitat is a great destination for a lot of finds; I'm always confident that they'll get to the right place and improve the quality of people's homes. Best of all, many ReStores offer free pickup, depending on the item.

REUSE CENTERS

The Big Reuse is currently located in Brooklyn, New York, but planning to expand to other markets. It's a nonprofit outlet for salvaged and surplus building materials, appliances, and furniture, accepted by donation and offered to the public and builders at a deep discount. Other, similar reuse centers have cropped up all around the country in recent years. A few of them are Rebuilding Exchange, in Chicago; the Loading Dock, in Baltimore; the Green Project, in New Orleans; the ReUse People of America, in Oakland; Architectural Salvage Warehouse, in Detroit; and the Resource Exchange, in Philadelphia.

Salvage Warehouses

There are also many for-profit businesses that sell salvaged items for building. These are typically more high-end than reuse centers, often used by architects and home designers, but they offer unique, valuable, and reusable materials at lower-than-expected prices and practice sustainability by saving perfectly good items from landfills. Some of these are Olde Good Things, in New York City and Los Angeles; Urban Remains, in Chicago; Eco Relics, in Jacksonville; Discount Home Warehouse, in Dallas; and Salvage Heaven, in Milwaukee. There is likely to be a salvage warehouse in your area and, if you have the right kind of dumpster find, it may offer you a great resell opportunity that you wouldn't get just from posting online.

Thrift Stores

They're seemingly everywhere, and they offer used goods at discount prices and take donations of all kinds of useful items. When it comes to thrift stores, these are the two biggies:

THE SALVATION ARMY

The Salvation Army describes itself as "an international movement" and "an evangelical part of the universal Christian Church." Most of us know it from its bell-ringing charity drives during the Christmas season—and for its ubiquitous Salvation Army Thrift Stores, which seem to exist in almost every town and city. Many of these stores offer free pickup of donations or specific hours when they take drop-offs, and all of them help the community by offering gently used goods at low prices.

GOODWILL

Like the Salvation Army, Goodwill is an organization devoted to community service that operates a vast network of community-based thrift stores that accept donations. In addition to offering gently used goods at low cost, the profits from Goodwill's storefronts also fund the organization's job-training and placement services. They operate career centers that can help with job search, résumé writing, rehab programs for work-related injuries, interview skills, and even getting a suit for that interview. They are one of my favorite organizations because they touch people in need at so many different pain points, helping them not only survive tough times but get back on their feet.

Buy Nothing

The Buy Nothing Project was founded by two friends in Bainbridge Island, Washington, in 2013. It has since grown into a global network of community-based groups in thirty countries. It promotes the "gift economy," that is, the giving—and receiving—of consumer goods and services as opposed to using cash. In the founders' words, Buy Nothing is about "creatively and collaboratively sharing the abundance around us." In addition to charities and reuse centers, this can be a great way to get a useful dumpster item that you don't need into the hands of someone who does (and get something useful back in return).

My favorite thing about the Buy Nothing program is that, in addition to preventing waste and helping folks in need, it builds a natural community where there wasn't one before. Most people who swap stuff on Buy Nothing will be back for more—either because they see how easy it is to save money and not buy new or because they like the feeling of exchanging with neighbors. Or both! Like diving, it becomes a habit and a lifestyle, good for both your wallet and the planet.

Removal Services

Don't have the brawn or the bandwidth to bring your donations to the thrift store, reuse center, or salvage yard?

Never fear. Those treasures you found don't need to sit in the backyard or basement. You can hire a service that offers pickup of items to be donated or recycled responsibly. Here are some notable examples.

JUNKLUGGERS

Junkluggers are pioneers in sustainable junk removal and provide service to communities from coast to coast. Their mission: To make a positive impact by hauling away junk from customers' homes or businesses and "either donating it to local nonprofits or recycling it in an eco-friendly way."

Other sustainable junk/donation pickup services include GreenDrop (Philadelphia/mid-Atlantic), Donation Nation (Washington, DC; Maryland; and Virginia), JUNK Relief (Chicago), Eco-Dumpster (San Francisco), Angel's Junk Removal (Seattle), and many more.

And remember, if you end up with electronics, batteries, or other nonbiodegradable items you don't need and can't sell, take them to your local recycling place to be correctly disposed of. We definitely don't need any more of those in the landfill!

5

How to Sell It

In addition to finding usable items for my home and donating goods to charities, I've discovered reselling to be rewarding, both in terms of the happiness it gives me to make items available to others and as a fun way to make a little extra money. Let's be honest, a lot of us have had the fantasy of creating our own little shop somewhere, right? Thanks to dumpster diving, your inventory is free (aside from your efforts at finding worthwhile items and making sure they're in good shape or reconditioning them). Two categories that are great for reselling are electronics and clothing/fashion.

Electronics

eBay

In my experience, electronics sell better than clothes—people are always looking for gently used laptops, phones, video-game consoles, gadgets, and, in some cases, vintage stereo and home-theater equipment. A good place to go with this kind of merchandise is eBay. (My own income stream from selling on eBay consists mostly of electronics sales.) To determine a good price for an item, I usually search eBay to see what similar items are selling for and then set a price accordingly. (The website Swappa, a user-to-user marketplace for "gently used" technology, also offers a way to check the going rate for any device.) eBay does allow buyers to effectively negotiate prices, using a bidding system. (There is also the Buy It Now option, which allows you to place a set price on any given item.)

Amazon

Amazon offers a trade-in program for its devices, including Kindles, Echo, home-security paraphernalia, tablets, wireless routers—the kind of stuff that often lands in a dumpster when the user buys the next great thing that comes along. You just enter some details about the item (storage, screen size, and so on) and describe its condition. Amazon then makes an offer, and, if you accept it, they give you a shipping label and you send it away. Your Amazon account is then credited with an eGift card.

Apple

Much like Amazon, Apple's trade-in program takes its devices—iPhones, iPads, iPods, Apple Watches, and MacBooks—and offers either credit toward the next purchase or an Apple Gift Card. (And if the device doesn't rate any credit, they'll recycle it.) Not many people seem to know this, but Apple's trade-in program also accepts Android devices from Samsung and Google. Believe me, there is very often an iPhone in a dumpster.

Best Buy

Best Buy also has a great trade-in program, covering any number of gadgets, devices, and appliances. You just fill out an online questionnaire and get a quote. The offer amount is awarded as a Best Buy gift card.

Clothing/Fashion

eBay

Yes, eBay blazed the trail for every at-home vendor, and it's what I use. For me there's a comfort level, and I can sell anything I want. The platform has been great for both buyers and sellers of vintage clothing, whether they be fashionistas or folks just looking for a winter coat.

Etsy

These guys rewrote the book on vintage and craft online. If you're really into clothes, you may want to sell your well-curated line of found fashions here, creating your own online boutique. On the upside, seller fees are on the lower end of the spectrum, and Etsy shoppers tend to be adventurous and knowledgeable.

Facebook Marketplace

If you're on Facebook all the time anyway, selling on Facebook Marketplace could make a lot of sense. There are no seller fees or taxes, and you can easily share your listings to your network.

Amazon Marketplace

You can sell used items on Amazon by becoming an Amazon seller. Amazon offers a range of seller programs, from individual to small business—you can find out more on sell.amazon.com. The process is pretty streamlined, but there are fees—either a per-item fee of ninety-nine cents or a flat monthly fee of $39.99—so it's a good idea to get a sense of how many salable things you're picking up month to month to make sure you'll be covering your costs and to help you pick the right plan. With secondhand items, you'll also need to scan the Condition Guidelines for each category of merchandise you plan to sell. (There are some items that are considered restricted for sale, like alcohol, groceries, topical creams, and dietary supplements—Amazon provides a full list.) The upside is that Amazon is easily the largest marketplace in the world, so it may be the best way for buyers to find you. And if you plan on branching out, at any point, from reselling found goods to making and selling your own (clothes, toys, etc.), it definitely helps to have an Amazon storefront.

The RealReal

The RealReal is a high-end consignment sale site for your major discoveries. It bills itself as "the world's largest online marketplace for authenticated, consigned luxury goods." The focus is on designer clothing, shoes, and accessories, but the RealReal also accepts fine jewelry, decorative pieces, and even fine art for sale on its site. The upside for the seller is that you can present your quality item to a discerning audience; for the buyer, a necklace or watch can be purchased for below the estimated retail price. The RealReal offers free pickup in select metropolitan areas, and now even has a few physical storefronts in places like LA and New York, so you can get a sense of their stuff and see whether any of your recent finds would fit.

Angel Williams

Will You Shop Thrifty for Your Next Coat?

Growing up in Chicago, I know what it's like to be COLD, and trust me, it's no fun. But it's a lot more bearable when you can look fabulous. 👄 👄 Now that we're full swing into January, it's time to move on to winter fashion and that starts with probably the most important thing you'll need this season—outerwear.

I'm totally obsessed with coats!! I guess you can call it my chic obsession. However, the only way I'm able to satisfy my obsession is to shop thrift!! Yes!! I scored some fly winter coats for a fraction of the cost from my local #goodwill and #salvationarmy.

Winter isn't going anywhere anytime soon. Oh well. The least I can do is dress in style and inspire you to do the same. 👄

How to Wear Sequins

Do you have a sequined piece in your wardrobe? If not, I challenge you to incorporate some diva-level sparkle into your wardrobe. Trust me, you'll thank me later.

I'll admit I was anti-sequin for a long time. It wasn't until I found this blue beauty at my local Goodwill that this changed. I wanted to see how it would look on me. I have to admit, this eco-fashion statement took my wardrobe to an entirely different style level, and I'm loving it!

So "go on honey, take a chance!" (If you're a *Coming to America* movie fan you'll know that line all too well!) Get you some dazzle and WERK!!

But remember the rules of sequin wearing:

- Be CONFIDENT when you're wearing sequins; confidence makes the sparkle look FLAWLESS!
- The sequins are the star of the show. You'll need to balance it with a basic bottom.

- Again, sequins are the star of the show! Over-accessorizing will not give that beautiful piece the attention it DESERVES . . . OKAY?! Now get out there and SLAY!!

Depop

This selling platform, described as a mix of eBay and Instagram, has gotten very popular very quickly, especially with teenagers. It's a good place to sell streetwear, sneakers, and other stuff kids love. I always check Depop if I find something cool my kids don't want to keep for themselves. Unlike some resale platforms, it's also very smartphone-friendly, which is nice if you're on the go a lot like me!

Craigslist

Yes, it's still around, and your buyers generally come to you.

Rehabbing Furniture

I'm not super handy myself, although I did successfully re-purpose a cube organizer that was completely destroyed, and I refurbished the desk in my prayer room. But I generally stick to pieces that are still in good shape and just apply a fresh coat of paint. And I've found plenty—for instance, I decided to redecorate my upstairs sitting area using ALL DUMPSTER-DIVING FINDS!! I'm amazed at how GREAT it turned out! If you can believe it, the small sitting chair, wall decor, picture frames, gold shelves, rug, and shelf decor ALL CAME FROM THE "TRASH"!!

But if true furniture rehab is something you're interested in, then diving can be a gold mine. Amazon has plenty
of books to help or check out shows like *Flea Market Flip*
on HGTV, or the YouTube channels I mentioned earlier.
One of the nice things about furniture rehab is it doesn't
have to be expensive (except for the value of your time, of

My upstairs sitting room is a space I'm extremely proud of
because I managed to decorate the entire room with
items I found sitting outside as "trash."

course). Assuming what you're picking up aren't priceless antiques, you can get wood and other materials cheaply from a lumberyard or overstock site. Favorite pieces that are always in demand include coffee tables, kitchen chairs, bookshelves, and desks. Especially now that so many people are working (and going to school) at home, a good desk is a real find.

Whether fashioning a new leg or two for a chair or refinishing a coffee table into the perfect living room accent piece, you can seriously improve the resale value of furniture—or its function in your own home—with just a little elbow grease. (Alternatively, you can use your talent to make something beautiful for someone who really needs it.)

If you need inspiration, there's always the internet:

- goodhousekeeping.com
- realsimple.com
- apartmenttherapy.com
- housebeautiful.com
- pinterest.com

As always, my followers have great ideas on repurposing and upcycling. Nancy ML says:

> I was given a baby bed years ago and never did anything with it, actually forgot about it. 😂 But I recently turned part of it into a full-size headboard for my daughter. I painted it brown and black, and it turned out fabulous!!

And Maureen Lucas reminds us that if you're buying new linens or towels:

> You can donate old towels and blankets to your local animal shelters. The animals would appreciate not having to sleep on a cold floor all the time.

Commenter SouperTrouper notes how different the recycling rules are from place to place—part of the reason dumpster diving has gotten so popular!

> I've always loved your videos. Yet I cannot understand the recycling there! Here in NYC, we have to empty and clean all our recyclables! You cannot just throw away unopened stuff. Ticketing is very widespread. Glad you are there to give perfectly good items a good home. Imagine if more people did that; we'd have less junk on earth.

Inspiration:
Helping Where You Live

Communities have always rallied around one another, but this year we saw it happen in so many creative new ways. One of my favorites was when small distilleries and breweries started making hand sanitizer instead of beer or liquor. Some repurposed their own equipment; others recycled thousands of gallons of beer that had been ordered by local bars and grocery stores but suddenly couldn't be delivered. More than eight hundred distilleries and breweries across the country participated, from Harpoon in Boston to Rogue Ales in Oregon and SanTan in Arizona.

Some of the earliest batches were delivered to YMCAs and emergency childcare centers for essential workers. So many people, especially in the food world, have lost their jobs or businesses this year—especially in the early days, when it was unclear whether restaurants and bars would ever be able to reopen. That's plenty to worry about! Yet they turned their attention to the bigger problem, to the people even more vulnerable than them.

Likewise, many restaurants and grocery stores, rather than throwing away the day's surplus food after closing (which, as many dumpster divers know, is mostly perfectly good) began donating it to shelters and food banks instead or setting up community refrigerators where people could take what they needed (or donate their own).

Gestures like these acknowledge that communities are interdependent; businesses need customers, people need good food and places to gather, and the local economy needs both. We all have a favorite restaurant, store, bar, or theater that's closed since COVID began, and most won't be back. But the best way to mourn them is to try to support the ones that remain, especially those that give back to their neighbors in need. And remember, when you're foraging for your own stuff, it's worth picking up things you may not personally need but that are in good condition. Someone can always use them. As follower Theresa Wilson told me a few months ago:

> You found some good stuff on this dumpster-diving adventure, Angel. My mom says the United States acts like a throwaway society. A person doesn't want clothing, they throw it away instead of giving it to someone else. Stores and boutiques have items zero out of inventory, they throw it away instead of donating it. I agree with her that many people and companies don't think to just leave it on the curb for someone to take home.

Finding Your Treasure

Or Tobi Kopachik-Murrell, one month ago:

The best part of the video for me is hearing the joy of giving in your voice. Helping others during hard times and the joy you find in your job brings happy tears to my eyes. You are changing lives around you. With your donations to others and your videos. May the heavens bless you and the spirits protect you on your adventures.

6

Learning the Lessons

When I first started diving, my family didn't know what to think. "When you started diving," my husband, Antwan, told me, "it felt crazy for you to be out the door early in the morning. And the only thing you wanted to talk about was dumpster diving. Any time we went somewhere, you felt compelled to peek into people's garbage." (Okay, I'll admit this is true.)

Our older kids thought diving was "gross." They were embarrassed by it—why on earth was their mom going through people's garbage? But they changed their tune when I started bringing home electronic games, phone accessories, toys, and barely worn (or never-worn) designer clothes.

To be honest, one of the most gratifying aspects of diving is seeing my children's faces light up when I give them something they really like. At the end of the day, it doesn't matter where it came from.

Once my family realized I was on a mission, they accepted diving as an important part of our lives. When I started spreading the word through YouTube and getting a positive response, they could see how my diving was impacting other people, improving lives. They realized it gave me a sense of purpose and fulfillment, one that they could share in. The kids learned that giving back is an important aspect of what I do—and something for all of us to do. My vlogging also led to my diving becoming more of a family affair. My kids are directly or indirectly involved in my videos: I'll mention them, or I'll find something for them, or they're caught on camera. (Sometimes they get stuck holding the camera!) Samuel has been the star of many of my latest videos, with his chubby cheeks and sweet patience. In fact, we recently started a new family channel so they can be even more involved! As one kind follower said:

Hi Angel 😊 I enjoy your channel. I have been watching for some time now, but never commented until I caught the tail end of your Thanksgiving Live Day stream. You have a lovely family! Samuel just makes me want to kiss those little cheeks through my screen. 😙 Thank you for sharing your Family and Thanksgiving with the world. God bless you and your family. 😇😇

I once pranked Taylor to try to get her to change her mind about dumpster diving. In my video called "She Hated Dumpster Diving Until . . ." from March 2017, I planted an outfit on top of a recycling bin and asked her to go check it out. She kind of suspected I was pranking her, but I wanted her to understand that many of the items I find are in equally good condition. As I explained, "Hopefully I've changed my daughter's thought process on the art of dumpster diving. If we keep our minds closed to opening up and doing different things, we will always be left behind. Because we are forever evolving, this world is forever changing, so as a mother I want to open my children's minds to doing all kinds of new, interesting, fun things—and hopefully I succeeded in doing that today."

Hailey recently joined me on a livestream where I was opening some boxes my followers had sent me. There was a Jenga set in perfect condition, a Kohler Bluetooth showerhead, and a full box set of Godzilla movies still in their wrapping. And then she giggled at Mommy when I opened an absolutely gorgeous box of clothes and shoes—my favorite! Most of them looked practically unworn, including a fabulous pair of cowboy boots, some studded suede boots, beautiful blue low-heeled sandals, and amazing gold mesh heels. There was also a box of stunning jewelry, including a pearl rope necklace and a woven gold one. Hailey knows I get almost as excited about shoes as she does about all the toys I bring home! (Of course, I picked a few favorites for myself, and the rest I'll share with family and friends.)

But just so it doesn't seem like I'm putting words in their mouths, I'll let them speak for themselves!

Antwan: "When she started doing the social media, I could see her growing as a person and feeling that she's impacting other people. That's when I started to see the benefits of diving. At first it was a hobby, now it's her FT job and vocation." And in the last year or so, Antwan has joined me more and more. My audience loves to see us together, and we can cover twice as much ground in half the time!

Josiah: "When I was six or seven, I didn't know what dumpster diving was—at first I was like, 'Why are you going thru the garbage?' But once you found that Xbox, that blew my mind, I thought, 'This cannot be true. Why would someone throw out an expensive item like that?' Diving has changed our lives."

Taylor: "I thought dumpster diving was a little crazy at first. I didn't really like it. I thought it was dirty. I guess I was a little embarrassed too. My impression has changed—it's hard to believe the name-brand stuff and gift cards you've found, some with a lot of money still on them. Now I'm more comfortable going with you, I can see how it helps others; I see the response on YouTube. I think anyone can try diving without being ashamed."

Hailey: "Sometimes I'm worried when you go out early by yourself, but I love the stuffed animals and toys you've brought home for me!"

Most important, because they hear me talking about the benefits of diving and can see in practical ways how it's changed our lives, my kids have become more aware of the larger questions diving addresses, ones we should all be asking ourselves, about consumption and conservation, recycling and reuse. Now, they wouldn't think of just throwing something away when they outgrow it or get a newer version.

Let's ask ourselves some of these questions now. It's always good to take a hard look at our habits.

- What was the last item you carelessly discarded? What recent items would you admit were wasteful in the end—like a jacket you ended up wearing once or a gadget the kids used a few times before moving on?

- Think about your next "must-have" item. Does it really need to be purchased new in a store or online? As my kids get older, I realize how silly it is to feel compelled to buy them the latest toy, game, or device that they may or may not end up loving as much as they think they will. We know that, with kids, today's hot item will quickly become yesterday's fad. Doesn't it track that, as adults, we also "outgrow" our immediate needs?

I believe we should all be more aware that our economy relies on a never-ending consumer cycle. Marketers urge us to buy things we don't really want or need. Some of us even go into debt to buy them. And, after that, when we throw them away, they wind up burdening our landfills. Sure, it's fun to live in a consumer culture. But the fact that so many useful things wind up in dumpsters is in itself a comment on the wastefulness of it.

- Does the idea of "living frugally" have a positive or negative connotation for you? To me, being frugal means being mindful— not just of money spent but of our larger impact on the environment. It's not just individuals who are wasteful but manufacturers, food suppliers, and businesses. If we bought less, they'd produce less. Join me in being a proud #frugalnista. My motto: I have the ability to turn nothing into something (and make money doing it)!

- Is the health of our environment important to you? I truly believe that by diving, in my small way, I'm helping to lower my environmental impact. Again, one item rescued from the dumpster is one less item in a landfill. These days every bit counts.

- How do you feel about the issue of income inequality, which has become a major conversation in our times? If some of the biggest problems we have in our society involve consumer waste and people in need, dumpster diving can address the first and help with the second. Today's dumpster diver may be a student, a recent graduate, a seeker of self-sufficiency, or an educated professional looking to conserve resources, live within their means, and give back.

- When was the last time you made a donation or gave back to your community in

some way? It's important for us all to realize the huge power of charity and giving
back. You don't have to be Bill Gates to do
it. We can all help in whatever small way
works best for us. It's my hope that I'm
changing your mind about dumpster diving one video at a time.

My followers always have great insight on paying it
forward too.

As Frugalnista commented long ago, when the girls
were little:

It's so good to give back, and for you to share that experience with your daughters is teaching another generation
to be compassionate. Thanks for sharing!

And as Angie Framke, who works for Goodwill, points
out:

Things are different now. Donations are coming in more
than ever, and everything has to be quarantined at the
warehouse before it can be sold, and the warehouse got
full fast! We had to cut back donation hours and some
days close, but that doesn't stop people from dumping,
and that's what gets thrown away. It is a shame.

Follower Catherine Thompson reminds us that there are also ways to maximize our impact by partnering with stores and companies, especially at certain times of the year:

> If you donate an old coat to Burlington Coat Factory they give a homeless person a new coat, so it's a really good deal. You might wanna look into that. I'm not sure if they do it all the time but it is something they do.

Inspiration:
Gowns4Good

The COVID-19 pandemic has birthed so many ingenious ways of giving back, especially in sustainable ways. You may have heard of Nathaniel Moore, a recent graduate of the business school at the University of Vermont, who decided early in 2020 that rather than throw away his graduation gown he would turn it into PPE for healthcare workers on the front lines. He eventually lunched a non-profit called Gowns4Good, encouraging other graduates to do the same with theirs. He then sends them a logo they can attach to their caps (which can't be upcycled), advertising the cause.

This story inspired me because so much has been up in the air this year, especially for our children. Mine have toggled back and forth between remote learning and hybrid school. And it's still unclear what the next school year will look like. For seniors and college students especially, it's devastating to miss out on the rituals and celebrations that come with milestones like graduation. But it's important not to get too deep into our own heads and remember how much suffering there is elsewhere—and to do what we can to help.

To date, Nathaniel has collected over fifteen thousand gowns, though many more have been requested by hospitals. As we look toward the end of the COVID crisis, we should remember that there will always be another health emergency on the horizon, and we need to support the folks who take care of us all. It's never too early to learn this lesson, so it's great to see today's young adults taking it to heart.

7

Family Business

Now that the Williamses are an official family business, we love to spread the word to friends and family! And sometimes getting new folks into the business pays off quick! One Monday this past January I went to a new location recommended by my brother-in-law Thein, who just got into diving himself (after hearing me talk about it so much). It was a phenomenal dive—I found a bunch of glassware, some of which was perfect for our pantry, and some of which I donated to a homeless shelter after cleaning and sterilizing it. I also found a brand-new pair of slippers, some great clothes for Josiah, and a huge haul of other clothes, good as new, for donation; a beautiful leather wallet, a movie theater gift card, and lots more. Best of all, I found a GoPro that I've used ever since to record my adventures!

Given that the clothes were mostly brand new, here's a plea: If you ever have a big bag of good clothes you don't need, please don't throw them in the dumpster! Chances are I won't be coming by you that day, so they'll end up in the landfill when they could be keeping someone warm. Clothes are the easiest thing to donate. If you don't have a Goodwill or a Salvation Army near you, look for clothing drives, homeless shelters, and even freestanding donation bins—they're everywhere these days.

And let's not forget about food, especially after the holidays. On my first dive of 2021, while telling my audience about an unfortunate experience where a pumpkin pie exploded all over my refrigerator, I found a huge bag of unopened, unexpired food—frozen spinach, fruit, you name it. At first I thought it must be a move-out, but then when I saw how much regular garbage they had too, I couldn't believe it! Please, please remember: If you have good food that you can't use, take it to a food bank or a community fridge!

High on Life

One thing I love about the blessing of my family is that they keep me high on life. We work together, play together, pray together, and, most important, take care of one another when we need to. I know my family will always be there for me when I need a little "me time" to keep from getting burned out. This past year we've all had plenty of time to be together, which is a blessing, but it also makes me think about the many people who weren't lucky enough to have as much support.

It's important to stay high on life, when there are so many other temptations out there, especially when we're feeling lonely or isolated. As my follower Amanda Marsh-Countryman recently said:

> Angel, I just wanted to say that I totally agree with what you said about being high off life . . . it is the only way to go! By our amazing God's graces, I've been clean off heroin, drugs and alcohol since 7/5/2013. I personally know how dark the other path can be and I just want to personally thank you for being a humble, ah-mazing Child of God that we can all look to for inspiration. Especially on YouTube when I have found them to be (in the dumpster-diving community) so few and far between. Thank you for sharing your life with us. Not only does it help us to know we're not alone, you help keep me motivated, grounded and walking strong in my faith. So much love from Washington State!

And follower Cindy Belcher made me laugh out loud when she commented: "I get high off a clean house."

There's never a dull moment when we come together as a family.

Our home is blessed with love and laughter,
an attribute that I can only thank God for.

As you all know by now, the rock of our family is my fabulous husband, Antwan. He's always been so support-ive of my diving, even when he wasn't sure where it was leading. I know it probably seemed a little strange at first, but he trusted me to follow my instincts, the same way we both trust in the Lord to show us the path. And look where we are now!

Back in 2014, I made a video called "My Secret to Suc-cess" that got a huge response from my followers. I think it's because a lot of folks were curious about how I man-aged to dive every day and take care of my family too. Here's what I said—and it's all still true today (even though Sammy has replaced Hailey as the baby).

As a working mom, I have to do my part. I can't let my hus-band do everything. I'm an able body too. "Every woman is born with a crown, and every experience she endures is meant to be a jewel in her crown. As women, our inner work keeps our crown straight. Our willingness to do the work is the ONLY way we can claim or reclaim our throne."[1]

What is my secret to succeeding? My secret is prayer. Before I come out here every day, I make sure I say a prayer to God. I know all things come from him. Nobody will succeed without giving praises. I pray to not only allow me to find things but also to keep me safe.

Why do I go diving during the day and let people see

me? My answer to that is hustle. This is my work. Me selling on eBay, and taking nothing and turning it into something, is my talent, and I'm going to use it to the best of
my ability.

A lot of you all have asked, how do I work with a baby?
Sometimes it can be difficult. But if I bring Hailey, I make
sure I have lots of snacks and let her watch some screens.
When we come home, we clean ourselves up & then I
take her somewhere she enjoys going. I try to keep her
active, I don't keep her in the car all day. And let's not for-

get, everyone in a family has a role to play. The older kids can help out with the little ones when Mom and Dad have a lot to do. It's good practice!

This business is SO competitive. Don't be greedy. That's how you end up not succeeding. Being greedy—there's more than enough out here for everybody. There's more than enough money to go around. Don't be mean to your fellow divers, there's no need. We can all prosper out here. We need to learn how to love one another, help one another, and just be kind to one another.

The moral of the story, for me, is that when you believe in your partner, it pays off for both of you eventually. The secret to a successful marriage, as I've told my fans before, is being on one page, as one team, with one accord, serving one God. And when you're on the same page about how to raise your children, it makes everything a whole lot easier! Antwan is a natural teacher and spiritual leader, so I know he appreciates the lessons that diving is teaching our children as much as I do.

In that spirit, I encourage other folks to make diving a family affair too, as much as they can. To me, it's the same as children learning where their parents work, and where the household money comes from. That way they appreciate it more. That's especially true today, when so

My husband is someone I truly admire. I feel protected whenever he's around. I'm extremely grateful to have him by my side.

many things are paid for electronically or automatically; when they don't see the money, it makes it feel imaginary, which we all sure know it's not! Once kids are old enough to understand how much things cost, they can understand how Mom and Dad provide for them (and even help out).

Last year, on Earth Day, I posted this message: "For all the parents watching this, I encourage you to go out &

dumpster dive w/ your child and get any little item that can be reused or resalvaged and do family DIY projects. And remember when you're out there, stay safe. Do not go dd w/out your protective gear, and if you're a minor, do NOT go dumpster diving w/out your parents." If you have a Michaels or a JOANN near you, it's always worth checking their dumpsters for lots of free crafting materials! I once found a box of wreaths, a bunch of empty picture frames with mattes, some wire baskets, ribbon, and more. That's a whole winter afternoon of crafting with the kids, for free!

Earth Day is an important holiday for us, so I celebrate it with my followers every year, encouraging them to get out in the neighborhood and clean it up. As I always say, your Earth loves you, so you love it back by keeping it clean! That's another value that's best passed down from parents to children. The sooner they see how much it means to you, the more likely they are to act on that value for the rest of their lives.

I think about all this so much more now that my older two are almost grown. This is the world they will inherit, the one they have to help clean up if we want to keep living in it. So whether they choose to continue the family business when they're out on their own or not, I'm glad we've instilled in them an appreciation for the values that drive it. They'll carry these with them as they start their own families too.

Conclusion

I can't believe our journey together is almost over! But let me say again how grateful I am that you chose to spend your time reading *Finding Your Treasure* and getting to know more about me, my family, and my passion for dumpster diving. I hope it's opened your eyes to the possibilities of making something out of nothing. Even if you never actually search through a dumpster, I trust you're now aware of the sheer volume of quality items people throw out without a second thought. If I've shifted your mindset just a bit so that you reflect on your personal consumption habits or any preconceived notions you may have about secondhand goods, I'll consider it a win. Hopefully you agree with me that divers are doing the world a great service by helping to save our planet. We are the embodiment of the environmental movement's mission to "reduce, reuse, recycle." We collect discarded items to reuse them, repurpose them, or give them to someone who really needs them.

If you'd like more inspiration, I invite you to follow me on social media. I've dedicated my Instagram account to highlighting the joys of sustainable fashion. On IG, I showcase the incredible outfits I find while diving or inexpensively thrifting. I get so much joy and fulfillment out of repurposing clothes and demonstrating how to be comfortable in your own skin. (You can find me on IG at instagram .com/momtheebayer101.) One thing the eco-fashion community has taught me is that trends, as much as I love them, are the cause of so much wasted "fast fashion"— and, let's be honest, after the age of eighteen, you can only rock certain looks anyway! Thrifting has sharpened my eye for what will actually suit me and helped me separate well-made and lasting pieces from the stuff that will fall apart tomorrow. That's not only good for keeping clothes out of the landfill, it helps keep you looking fierce!

And, of course, I hope you'll visit me on YouTube at youtube.com/c/MomtheEbayer101 to watch me in action and virtually experience the thrill of the hunt! You'll also get to "meet" my growing community of amazing followers, who hail from around the world.

I know this past year has been especially hard, an unprecedented challenge for all of us. I pray you stay safe and healthy, and that we see the end of this once-in-a-life-time pandemic in 2021. But even during this time when our choices seem so reduced, I believe we're still lucky to have choices that can make an impact on our world over and over, every day. I encourage you to choose a mindful, purposeful way of life over the consumer culture that has grown unchecked for years. Because even after this pandemic is over, there will always be another crisis waiting in the wings: medical, environmental, social. And the best way to mitigate the damage in an unpredictable future is to think hard about the way we live today.

The other thing that's become clear over the past year is how many folks in America are living on the edge—of poverty, of homelessness, of despair. I hope some of the people and organizations I've talked about in this book who are responding to the extraordinary needs of their communities during this crisis have made you think about what you might be able to do to help. I know not everyone has extra money or things they can spare, but there's always a way to donate your time, if you can find a few hours. Volunteering in a soup kitchen or a shelter, reading to kids at an after-school program, or delivering meals to the sick and elderly are all crucial services that people depend on now more than ever. And, of course, you can always regularly check in on the people around you who might be feeling low—friends, family, neighbors. It's amazing what a difference it can make just to hear a friendly voice and to know someone is thinking about you. It's all in the spirit of giving back.

The great thing is, I've found that giving back does almost as much good for you as it does for the world. You carry yourself differently, hold your head higher, and approach people with more openness than when you're wrapped up in your own problems. That energy then radiates out and lifts up others too. One of my followers recently asked me about how to gain self-confidence. (As she correctly perceived, diving takes a lot of self-confidence, especially when you're filming yourself exploring other people's garbage and posting it for the world to see!) I'd like to share my answer, and I hope it shows why I feel so strongly about giving diving—or any new activity—a try:

I've learned over the years that when you're not happy with yourself, everything around you will be dark. You can't be happy or joyful for other people's sake; ironically, it only ends up weighing you down. If you're not happy with who you are as a person today, I recommend setting aside some time each and every day to work on it—to sit with your feelings, pinpoint where they're coming from, and try new things that might help you grow. I know that when I was in my own place of darkness, it wasn't that I was miserable with my life or hated being around people. No, my darkness came from wanting to change the way I acted *toward* others. So what did I do? Of course, I prayed to the Lord. I got honest feedback from people around me, including friends and family, about how they viewed me. I read books by teachers I trust, and I made it my business to do at least one thing every day that would help or show love to another person. Pretty soon it became a habit, an integral part of the person you see in front of you. And it's incredible how much better I felt—and how much my relationships with everyone around me improved.

I know my followers agree, and have seen the same incredible transformation happen in their own lives. As commenter Louise Nilsson wrote just a couple of months ago:

> I find that you get confidence when doing something you love and feel you are good at. Fear and anxiety can cause lack of confidence. Being in a situation that is unfamiliar can cause bad confidence. So it is good to get used to different and new things. I also think that it is a false sense of confidence when people only care about what they look like on the outside. It has to come from within, from what we do and also from not thinking about ourselves so much.

I do believe confidence comes from within—specifically from being deeply, sincerely happy with yourself. It's not something somebody can give you. It starts, in my opinion, with self-love. When you love yourself, you can be confident in everything you do.

This is a lesson I think it's also important to teach our children, especially since they live so much of their lives now on social media (and you know I love social media, of course!). But it's different for grown-ups—we've had a little more time to develop a thicker skin. Kids today are constantly dealing with unwanted opinions about their hair, clothes, friends, everything. It's enough to make them lose heart before they even start. So the best lesson we can teach them is that they're enough—that their worth isn't tied up in any flashy thing they own but in the acts of kindness they perform. And there's no better way to teach this than to model it yourself.

If you're looking to change the way you feel, I encourage you to try to explore things that you normally wouldn't do. If you consciously examine your everyday reactions and habits, you'll learn which ones you want to work on. I honestly believe self-examination is the beginning of self-love, self-belief, and self-confidence. One of the things I try to encourage in my social media posts is to acknowledge who you are and to be bold in your gift, whatever it is. Don't be afraid to be who God created you to be.

One follower named Barbie wrote, in response to one of my fashion finds, "You are teaching us—or at least me—how to get out of the box." Marcella Orosco wrote, "The way you bless others, it's all coming back to you three-fold. You're such an amazing inspiration to me and to so many more fans. May God continue to bless you and your family." And Melanie Lippincott left this amazing message: "You are an incredible woman. Your peace, love, joy and contentment are contagious. You are the salt and light, exactly like Christians are called to be. You face your struggles like everyone else, after all we learn from failures not success. Keep it up chica. Don't grow weary in well doing. You are validated and acknowledged!"

And, of course, you've got to have confidence in the Lord. With God all things are possible. Without Him, you can't do anything. With His love will come confidence, joy, love, and peace—all of it goes hand in hand.

Stay humble,

Angel

Acknowledgments

I give all honor and praise to the true and living God, the Heavenly Father and his son Jesus, through whom all blessings flow! None of this would be possible without Him.

My phenomenal husband, Antwan—my life partner, whom I love dearly, and whose love for me goes beyond words. As we continue to do this thing called life, I'm eager to experience all that the Lord has in store for us— together on one accord!! Thank you for your support with this book, the countless hours of motivational speeches at two o'clock in the morning, the forehead kisses, and, most important, the silent prayers you lifted up to God on my behalf.

To my four wonderful children, Taylor, Josiah, Hailey, and Samuel; my nieces Journei, Autumn, Silver, and Prima;

my nephews Izaiah and DJ; and my goddaughter, Laila—I love you!

You're either too young to read this or simply don't have a desire to right now, but someday you will. When that day arrives, I want you to know this: All things are possible if you believe that you can! Never give up on your dreams, because you have what it takes to make a difference! And on another note . . .

I told ya dumpster diving wasn't that bad!! 😜

Ma & Daddy, there's no me without either of you! Life is a journey ready to be taken on and explored. This experience wasn't always easy, but I appreciate having you as parents to equip me for all of its realities. I love you both!

To my little/big brother Kyeran: Thank you for always making my stomach ache with laughter. There's never a dull moment with you around. I love you, little brother.

I also want to thank my wonderful sister, Angela. Anyone who knows me knows we're joined at the hip. I'm forever grateful for your continual love! Thank you for always being my number one supporter in everything I do. Thank you for always speaking life into me when I'm unable to see clearly. It's a blessing to have you as a sister. I love you.

In addition, a special thanks to my soul sister, Elana C., whose jokes, kindness, and affectionate spirit instantly drew me to her. You, my dear, are my best friend for life! In sixth grade we weren't too sure how far our friendship would go, but today it's safe to say it stood the test of time. There's nothing I've done that you haven't supported me on, and I'm truly grateful! I love you!

My amazing sister-in-law, Lisa, and my brothers-in-law Doug and Thein: All of you came into my life by marriage. From the moment we met, you never ceased to amaze me. Your advice and support will always have a special place in my heart. I love you!

Thank you to all the wonderful women in my life! Each of you has inspired me in many different ways. You know who you are—and I appreciate and love you all!

Then there's my beautiful online angels: Thank you so very much! Every time I turn on my camera I know one of you will be inspired. It's a constant reminder that my work isn't in vain. I appreciate your comments, emails, phone calls, and letters. I will forever remember you and be grateful for the outpouring of love you bestow upon me daily. I love you all!

As I always say, I have the best YouTube subscribers on the platform . . . period!!!

Finally, to all those who have impacted, influenced, and inspired my life in meaningful ways—and there are too many to name—words can't express my sincere appreciation for each of you! Thank you!

Notes

Introduction

1. "Living Paycheck to Paycheck Is a Way of Life for Majority of U.S. Workers, According to New CareerBuilder Survey," Career Builder, August 24, 2017, http://press.careerbuilder.com/2017-08-24-Living -Paycheck-to-Paycheck-is-a-Way-of-Life-for-Majority-of-U-S-Workers -According-to-New-CareerBuilder-Survey.

Chapter 1:
How Our Faith Drives Our Diving

1. Marilyn Binkley and Trevor Williams, "Reading Literacy in the United States: Findings from the IEA Reading Literacy Study," National Center for Education Statistics (NCES), June 17, 1996, https:// nces.ed.gov/pubsearch/pubsinfo.asp?pubid=96258.

Chapter 3:
How to Be Prepared (and Stay Safe!)

1. Julie Bosman, "Domestic Violence Calls Mount as Restrictions Linger: 'No One Can Leave,'" *New York Times*, May 15, 2020, https://www .nytimes.com/2020/05/15/us/domestic-violence-coronavirus.html.

Chapter 7:
Family Business

1. "Lisa Ray McCoy," *Iyanla, Fix My Life*, season 7, episode 705, November 28, 2020, https://www.oprah.com/own-iyanla-fix-my-life/lisaraye-mccoy.

About the Author

Angel Williams has a YouTube channel with over 140,000 subscribers, posting under the name Mom The Ebayer. She has an additional 7,000 followers on Instagram. With her husband and children, she runs a resale business for found goods on eBay.